我爱柠檬汁

I Love Lemonade

Gunter Pauli

[比] 冈特·鲍利 著

[哥伦] 凯瑟琳娜·巴赫 绘

朱　溪 译

上海远东出版社

丛书编委会

主　任：贾　峰

副主任：何家振　闫世东　郑立明

委　员：李原原　祝真旭　牛玲娟　梁雅丽　任泽林

　　　　王　岢　陈　卫　郑循如　吴建民　彭　勇

　　　　王梦雨　戴　虹　靳增江　孟　蝶　崔晓晓

特别感谢以下热心人士对童书工作的支持：

匡志强　方　芳　宋小华　解　东　厉　云　李　婧

刘　丹　熊彩虹　罗淑怡　旷　婉　杨　荣　刘学振

何圣霖　王必斗　潘林平　熊志强　廖清州　谭燕宁

王　征　白　纯　张林霞　寿颖慧　罗　佳　傅　俊

胡海朋　白永喆　韦小宏　李　杰　欧　亮

目录

Contents

一只橘凤蝶幼虫正享受一顿健康餐，狼吞虎咽地啃食着一棵小柠檬树的树叶。一只凤头鹦鹉飞落到树枝上，看着这只毛毛虫说：

　　"起先我还以为你是一坨新鲜的鸟粪，现在我才发现你原来是毛毛虫啊。"

An orange dog caterpillar is enjoying a healthy meal, devouring leaf after leaf of a young lemon tree. A cockatoo alights on a branch, and spotting the insect he says:
"At first, I thought you were a fresh bird dropping, but now I see you are a caterpillar."

一只橘凤蝶幼虫正享受美餐。

An orange dog caterpillar is enjoying a meal.

你等着，我还可以伪装成一条蛇呢！

Wait till you see me imitating a snake!

"好吧，我得承认这不是我第一次被错认为一团便便了。我甚至还被人称为'鸟粪'毛毛虫……不过你等着，我还可以伪装成一条蛇呢！"

　　"看上去像鸟粪，这伪装也太棒了！不过在你受到威胁的时候伸出分叉的舌头，让自己看起来像一条蛇，这更是巧妙又有效的防御呢。你的敌人肯定会大倒胃口的！"

"Well, I have to admit it is not the first time I have been mistaken for a piece of poop. I am even called 'the bird poop' caterpillar… but wait until you see me imitating a snake!"

"That's great camouflage, looking like bird poop. But what an ingenious and effective defence, showing a forked tongue that makes you look like a snake, when you feel threatened. Your enemies will certainly lose their appetite!"

"对，这能够让我安全地在这棵小柠檬树上尽享美味的树叶，直到我变成美丽的大蝴蝶。"

"其实我也很享受柠檬树为我提供的东西。我们凤头鹦鹉很喜欢健康水果，即便是酸的那种。实际上因为太酸了，只有极少数的鸟儿喜欢吃。这样一来，有兴趣吃的鸟儿越少，我能吃到的就更多了。"

"Yes, it allows me to safely feast on the tasty leaves of this young lemon tree – before I turn into a beautiful, big butterfly."

"I also enjoy what the lemon tree has to offer, you know. We cockatoos like the fruit, which is healthy – even if sour. So sour, in fact, that very few other birds will feast on it. Fewer birds interested, more for me."

我们凤头鹦鹉很喜欢健康水果，即便是酸的那种……

We cockatoos like the fruit, even if sour ...

加很多糖和水……

Adding a lot of sugar and water ...

"我知道人类会用柠檬加很多糖和水来做柠檬汁。"毛毛虫说。

"噢，什么时候人类才能学会戒掉这些糖呢？这只会抵消柠檬的益处啊。"

"正是如此！添加了那么多的糖会让骨头脆弱，还会导致体重增加。是的，我们都认为人类不应该在柠檬汁里加糖了。"

"你知道柠檬汁甚至能溶解岩石吗？"凤头鹦鹉问。

"I know people use lemons to make lemonade, by adding a lot of sugar and water," Caterpillar says.

"Oh, when will people learn to cut out all that sugar? It will only undo the goodness of the lemon."

"Exactly! So much added sugar weakens their bones and leads to weight gain. So, we agree, they should not add sugar to lemonade."

"Did you know that lemon juice could even dissolve rocks?" Cockatoo asks.

"你太夸张了！岩石那么难打碎，新鲜的柠檬汁怎么可能轻易做到？"

"柠檬汁的酸性很强，可以溶解坚硬的东西。现在我们来看看，人类还可以用这种一年四季都能生长的亮黄色水果做些什么吧？"

"举例来说，他们可以用柠檬汁来清洁铜和黄铜！"

"清洁？我以为柠檬只能用来吃喝呢！"

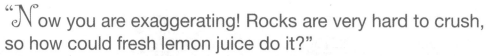

"Now you are exaggerating! Rocks are very hard to crush, so how could fresh lemon juice do it?"

"Lemon juice is very acidic, and can dissolve hard objects. Now let's see, what else could people do with this bright yellow fruit, that grows all year round?"

"They could, for example, use lemon juice to clean copper and brass!"

"Cleaning? And I thought lemons were only used for eating and drinking."

岩石那么难打碎……

Rocks are very hard to crush ...

柠檬皮可以使黄铜铃铛焕发光芒……

The peel can be used to shine brass bells ...

"嘿，在自然界里是没有废物的。除了利用柠檬汁以外，柠檬皮还可以使黄铜铃铛和铭牌焕发光芒。"

"毛毛虫，看来你是柠檬专家啊！"

"嗯，我们在柠檬树上居住了上千年，深知它们对世界意味着什么。我们见证了无数次错误尝试，也见证了成功。"

"Hey, in Nature nothing ever goes to waste. Apart from using the juice, the peel can also be used to give a great shine to brass bells and name plates."

"Caterpillar, you seem to be an expert on lemons!"

"Well, we have lived on lemon trees for thousands of years, so we know what they mean to the world. We have witnessed many trials and errors, as well as successes."

"我听说柠檬甚至能为电子表供电呢……"

"那要用到好多柠檬才行！还有，因为柠檬香味能够让人身心放松，所以柠檬叶很适合用来泡茶。另外，柠檬汁可以当作隐形墨水来写字。"

"谁会想用隐形墨水来写字呢？"凤头鹦鹉问。

"I heard that a lemon could even power a digital watch …"

"You may need quite a few lemons to pull off that trick! Anyway, lemon leaves are good for making tea, as the lemon aroma relaxes the mind. And when you write with lemon juice, you have an invisible ink."

"Who would want to write with invisible ink?" Cockatoo asks.

谁会想用隐形墨水来写字呢?

Who would want to write with invisible ink?

如果你感冒了，喝茶的时候最好加一点柠檬

Add lemon to your tea if you catch a cold

"应该是那些想要发送秘密消息，或者写情书的人吧。把纸放到火上烤一烤，或者用熨斗来熨一下，当纸张变热后，写的内容才能被看到。"

"这些柠檬真是奇妙啊。"

"我还可以说出很多呢。比如说，柠檬汁能防止切开的苹果和鳄梨变成棕色，还可以做嫩肉剂；如果你感冒了，喝茶的时候最好加一点柠檬。"

"Those who want to send secret messages, or perhaps love letters, where the writing only becomes visible once the paper is warmed, for instance by holding it over a flame, or ironing it."

"Fascinating things, these lemons."

"I could go on an on. For instance, lemon juice keeps cut apple and avocado from turning brown; and it is a meat tenderiser; and if you catch a cold, better add lots of lemon to your tea."

"你不该浪费时间只坐在树上！你应该去创业，建议人们充分利用柠檬来赚钱……"

"或者我自己做一名创业者，到街角去卖健康无糖的柠檬汁！"

……这仅仅是开始！……

"You should not be wasting your time sitting in a tree! You should start a business, advising people on the many uses of the lemon, and how they could make money…"

"Or I could rather become an entrepreneur myself – and start selling healthy sugar-free lemonade on a street corner!"

... AND IT HAS ONLY JUST BEGUN!...

……这仅仅是开始！……

… AND IT HAS ONLY JUST BEGUN! …

Did You Know?

你知道吗？

Lemons are a key component of preventive medicine. By eating one lemon a day, and walking regularly, you can reduce blood pressure. It also lowers the risk of suffering a stroke, especially in women. Eating lemon peel helps reduce weight.

柠檬在预防医学中有重要作用。每天吃一个柠檬并有规律地散步，可以降血压。柠檬还能降低罹患中风的风险，尤其对女性而言。吃柠檬皮有助于减肥。

The orange dog caterpillar (Papilio cresphontes Cramer) turns into the showy, black-and-yellow, giant swallowtail butterfly. The butterfly lays eggs on lemon leaves, and when the caterpillars hatch, they have a mottled grey, brown, cream and tan appearance that resembles fresh bird droppings.

橘凤蝶毛毛虫（美洲大芷凤蝶，Papilio cresphontes Cramer）会变成艳丽的黑黄色的巨型燕尾蝶。这种蝴蝶会在柠檬叶上产卵，当幼虫孵化时，它们呈斑驳的灰色、棕色、奶油色和棕褐色，外观类似于新鲜的鸟粪。

Lemon tree predators include (1) aphids: soft-bodied, pear-shaped insects, usually green, yellow, black or grey; (2) mealybugs: greyish, fuzzy-looking oval insects that attach to leaves and stems; (3) whiteflies: that settle under the leaves. All these insects suck sap.

柠檬树的天敌包括：（1）蚜虫：身体柔软的梨形昆虫，通常为绿色，黄色，黑色或灰色；（2）粉蚧：附着在叶子和茎上，灰色并且看起来覆有绒毛的椭圆形昆虫;（3）粉虱:在叶下定居。 所有这些昆虫都会吸食柠檬树的汁液。

Lemonade is one of the oldest commercialised drinks in the world. It was produced and exported from Cairo in the 11th century. In the 17th century, the lemonade makers of Paris were ordered to form a guild of their own.

柠檬水是世界上最古老的商业饮料之一。11世纪开罗就进行生产并对外出口。17世纪，巴黎的柠檬水生产商被命令成立行业协会。

Making and selling lemonade on a street corner can offer children their first taste of capitalism. Fizzy lemonade was industrialised in 1780, by the Swiss man Johann Jacob Schweppe, who used a compression pump to mass-produce fizzy lemonade. The brand is still selling strong.

在街角制作和出售柠檬水可以让孩子们有初次做生意的体验。柠檬气泡水由瑞士男子约翰·雅各布·史威普于1780年实现工业化生产，他使用压缩泵批量生产柠檬气泡水。现在这个品牌仍然畅销。

Lemons were a prestige fruit in the Roman Empire. Emperor Nero ate lemons every day, believing that the juice would neutralise poison. It was also used in the Roman era as a moth repellent. Lemonade arrived in Europe via the Ottoman Empire.

柠檬在罗马帝国时期是一种名贵水果。尼禄皇帝每天都吃柠檬，因为他相信柠檬汁可以中和毒素。在罗马时代它也被用作抗蛀剂。柠檬水在奥斯曼帝国时期到达欧洲。

The cockatoo is a parrot. Illegal trading and habitat loss have led to the decline of the species. The cockatoo has a loud voice, one that is unique to each bird, and used to alert each other to predators. The cockatoo also drums with sticks on a dead branch.

凤头鹦鹉是鹦鹉的一种。非法贸易和栖息地丧失导致该物种减少。凤头鹦鹉的声音响亮，每只鸟都有独特的声音，用来彼此间提醒避开天敌。凤头鹦鹉还会用树枝在枯枝上打鼓。

The cockatoo enjoys hanging upside-down. It has a preferred left or right claw, just like people are right or left handed. They prefer to learn to talk from other cockatoos rather than directly from a human.

凤头鹦鹉喜欢倒挂。它偏好使用左爪或右爪，就像人类有右撇子和左撇子一样。它们更喜欢学其他凤头鹦鹉说话，而不喜欢直接学人说话。

Would you mind looking like a piece of excrement, if it meant you would be safe from enemies?

如果你长得看起来像一坨便便，就意味着可以安全地避开敌人，你会介意吗？

Could you drink unsweetened lemonade, that is super-healthy, or is sugar needed?

你喝超级健康的不加糖的柠檬水，还是要加糖？

Do you like the idea of writing with invisible ink?

你喜欢用隐形墨水写字的想法吗？

Is selling lemonade on a street corner a good way to start a business?

在街头卖柠檬水是不是创业的一个好主意呢？

We are all aware of the health hazards of consuming too much sugar. So, it is time to research some drinks that are not sweetened. Find out who likes unsweetened lemon juice and who enjoys a spicy beetroot juice. Is there anyone who prefers a bitter taste? What are the health benefits of unsweetened drinks? Do a survey, and ask your friends and family members if they are aware of these benefits, and of the adverse effects of sugary drinks. Observe to see if their behaviour changes after you have shared the information with them.

我们都知道吃太多糖有害健康。是时候来研究一些不加糖的饮料了。看看谁喜欢不加糖的柠檬汁,谁享受带辣味的甜菜根汁。有喜欢苦味道的吗? 不加糖的饮料对健康有什么好处呢? 做个调查,问问亲朋好友是否了解不加糖饮料的益处,以及含糖饮料的坏处。观察他们的行为是否会有改变。

学科知识
Academic Knowledge

生物学	柠檬是酸橙和香橼的杂交品种；柠檬科包括橙子和橘子；巨尾凤蝶是北美最大的蝴蝶；毛毛虫的丫腺看起来像蛇舌；黄蜂是毛毛虫的主要天敌；雌性毛毛虫在决定产卵的位置时，会用触角来识别植物。凤头鹦鹉是鹦鹉科的一员，呈灰色。
化 学	一个柠檬含有30毫克维生素C；黄酮类是抗氧化剂；柠檬汁的pH值为2；柠檬酸和苹果酸的来源；巨大的黄尾蝴蝶通过植物中的挥发性化合物来识别其寄主植物；蛋氨酸是人体必需的一种氨基酸，同时也是防治毛毛虫的天然控制剂。多低的pH值才能溶解固体；柠檬汁如何延迟氧化并防止食物变色。
物 理	用三个柠檬做成电池能产生电压为2伏的电流；毛毛虫保护色的视觉效果掩饰了毛毛虫的身体轮廓，使其看起来像蛇；在海里变透明也是伪装。
工程学	伪装是指通过利用材料、颜色和光线来达到隐藏或使人炫目缭乱的目的，它消除了阴影和对立色；枪械射程和准确性的提高促进了军事伪装的发展；移动电话信号塔的伪装。
经济学	利基市场的重要性，将低需求、低竞争的细分市场视为目标市场；替代品在产品使用中的作用；将原材料用于多种用途来分散风险；差异化顾客需求（无糖柠檬水）并首先在本地市场（街角）销售的企业家风险较低；要取得成功，创新需要反复试错。
伦理学	使用杀虫剂消灭果园中的毛毛虫时，化学成分会增加环境负担，随之而来的是职业病危害和化学品残留对消费者造成影响。
历 史	在远洋航行中食用柠檬来增加维生素C的摄入，可防治坏血病；中世纪，人们用柠檬汁写密信。
地 理	柠檬起源于印度东北部的阿萨姆邦（中国和不丹之间）；橘凤蝶毛毛虫只能在美国佛罗里达州和南部诸州越冬。
数 学	计算销售利润的方法，毛利和净利是有差异的。
生活方式	糖的摄入量超出人体的代谢能力，导致糖尿病和肥胖，使得这种天然甜味剂成为公共健康问题；消费者对天然产品的偏好增加。
社会学	当局者迷，旁观者清。社交中保持谦虚：仅在适当的时候表现你的能力或分享你的见识。
心理学	嗜糖成瘾；当你倾向于只专注一个主题时，会忽略这一种用途之外的其他机会；香气如何影响我们的心情和状态；在共享的同时也要学会对部分事物保密。
系统论	由于缺乏霜冻，气候变化使得毛毛虫栖息地向北扩张；柠檬全身都是宝，充分利用柠檬的各个部分（包括果皮）。

情感智慧
Emotional Intelligence

凤头鹦鹉

凤头鹦鹉信心十足地分享了自己的观察结果（"你看起来像便便"），这个结论可能不会被普遍接受。他对毛毛虫成功伪装的能力不吝赞美。尽管只有很少的人意识到，但其实凤头鹦鹉通过再次确认柠檬树带来的益处创造了舒适区。因为竞争者较少，他满足于成为少数获得利好的种群之一。他坦率地发表了对糖的看法。柠檬酸足以溶解岩石这个事实令毛毛虫震惊，这一幕让他很享受。他承认毛毛虫知识渊博，但没有表现出羡慕或嫉妒。他乐于学习新的知识，并不惧承认自己的无知。稍后他还为毛毛虫提供了一些创业建议。

毛毛虫

毛毛虫有很强的自我意识，不觉得被凤头鹦鹉的言论所冒犯。她承认这不是第一次被误认为鸟粪了。在交谈中，她俨然是一位柠檬专家。关于分享糖对人体的影响，她展示了自己知识渊博的一面。她还提出了更多见解，列举了柠檬的用途，包括隐形墨水，这一切征服了凤头鹦鹉。受到凤头鹦鹉的启发，她透露了想创业的梦想。她没有吹嘘自己能成为最大、最美丽的一只蝴蝶，这表明她很谦逊。

艺术
The Arts

让我们把保护色融入艺术活动中。我们来画一幅森林里的一群鹦鹉。先用彩色墨水画树木、树枝和树叶。等干燥后，用柠檬汁当墨水，画出鹦鹉的轮廓。当图画干燥后，问问亲朋好友看到了什么，他们能不能在树叶之间发现鹦鹉。接着在成人的指导下，点燃蜡烛，把纸放在火焰上去烤，小心不要烧着。随着鹦鹉显形，注意看旁人的表情！（提示：也可以用热的熨斗。）

思维拓展
Systems: Making the Connections

　　自然界中没有浪费，因为所有生物都受益于其他生物的废弃物，并且没有不被利用的剩余物。如此，一切都有助于建立和延续可以维持其进化路径的成功生态系统。在自然界中总是通过合作来创建系统，使每个个体都有其特定的小生境，并保证所有个体都有合理的生存机会。通过研究其他生物如何促进健康的生态系统，我们将学到很多。分享这些知识也很重要，不是为了表现自我和打动别人，而是为了谦卑地努力合作，帮助他人以拥有更好的机会应对挑战。不同的物种如橘凤蝶毛毛虫会以独特的方式进化，以确保自身的生存以及对成功的、可以延续生命周期的生态系统作出持续贡献。在现代化时代，我们的专业化程度过高，以至于往往只关心人类的事务线，而以系统性方式为所有物种创造价值的重要性却被忽略——就像柠檬树，不仅把叶子提供给毛毛虫，把果实给凤头鹦鹉，也为我们提供叶子、油和果实。在一个所有物种都为成功做出贡献的生态系统中，其效率水平超过了单个做出贡献的生物体的效率水平。

动手能力
Capacity to Implement

　　想在街角摆个柠檬水摊吗？先完成两件重要的事：一是有设摊许可证，二是清水煮沸后才可使用。仅用有机种植的柠檬，这会增加柠檬水的吸引力。现在邀请朋友和邻居来品尝。你可以做一番市场调查，问问顾客是不是喜欢。注意顾客的反馈，并适当调整柠檬水的酸甜度。准确计算成本，看看怎样才能盈利。请记住，包括沃伦·巴菲特在内的许多企业家都是从销售饮料起步的。

故事灵感来自

This Fable Is Inspired by

阿曼达·福尔曼
Amanda Foreman

阿曼达·福尔曼生于伦敦。曾就读于莎拉·劳伦斯学院、哥伦比亚大学和牛津大学玛格丽特夫人学堂。1998 年在牛津大学获得 18 世纪英国史博士学位。阿曼达目前是《华尔街日报》"从历史上来说"双周专栏作家和利物浦大学历史系荣誉资深高级研究员。她是《德文郡公爵夫人乔治娜》的作者，该书连续数月居国际畅销书榜首，还获得多个奖项的提名，并于 1999 年获得惠特布雷德最佳传记奖。该书还诞生了一部电视纪录片，一部广播剧（朱迪·丹奇主演）和一部获奥斯卡奖的电影《公爵夫人》（凯拉·奈特莉、拉尔夫·菲恩斯等主演）。阿曼达的著作《燃烧的世界：分裂的两个民族的史诗》获得弗莱彻·普拉特南北战争史奖。作为一名历史学家，她撰写了《从历史上来说：柠檬简史》，她在这篇文章中讲述了柠檬水作为有史以来最具有历史意义的饮料的制造方法。

图书在版编目（CIP）数据

冈特生态童书.第七辑：全36册：汉英对照 /
（比）冈特·鲍利著；（哥伦）凯瑟琳娜·巴赫绘；
何家振等译.—上海：上海远东出版社，2020
ISBN 978-7-5476-1671-0

Ⅰ.①冈… Ⅱ.①冈… ②凯… ③何… Ⅲ.①生态
环境－环境保护－儿童读物－汉英 Ⅳ.①X171.1-49

中国版本图书馆CIP数据核字（2020）第236911号

策　　划　张　蓉
责任编辑　程云琦
封面设计　魏　来　李　廉

冈特生态童书
我爱柠檬汁
[比]冈特·鲍利　著
[哥伦]凯瑟琳娜·巴赫　绘

朱　溪　译

记得要和身边的小朋友分享环保知识哦！
八喜冰淇淋祝你成为环保小使者！